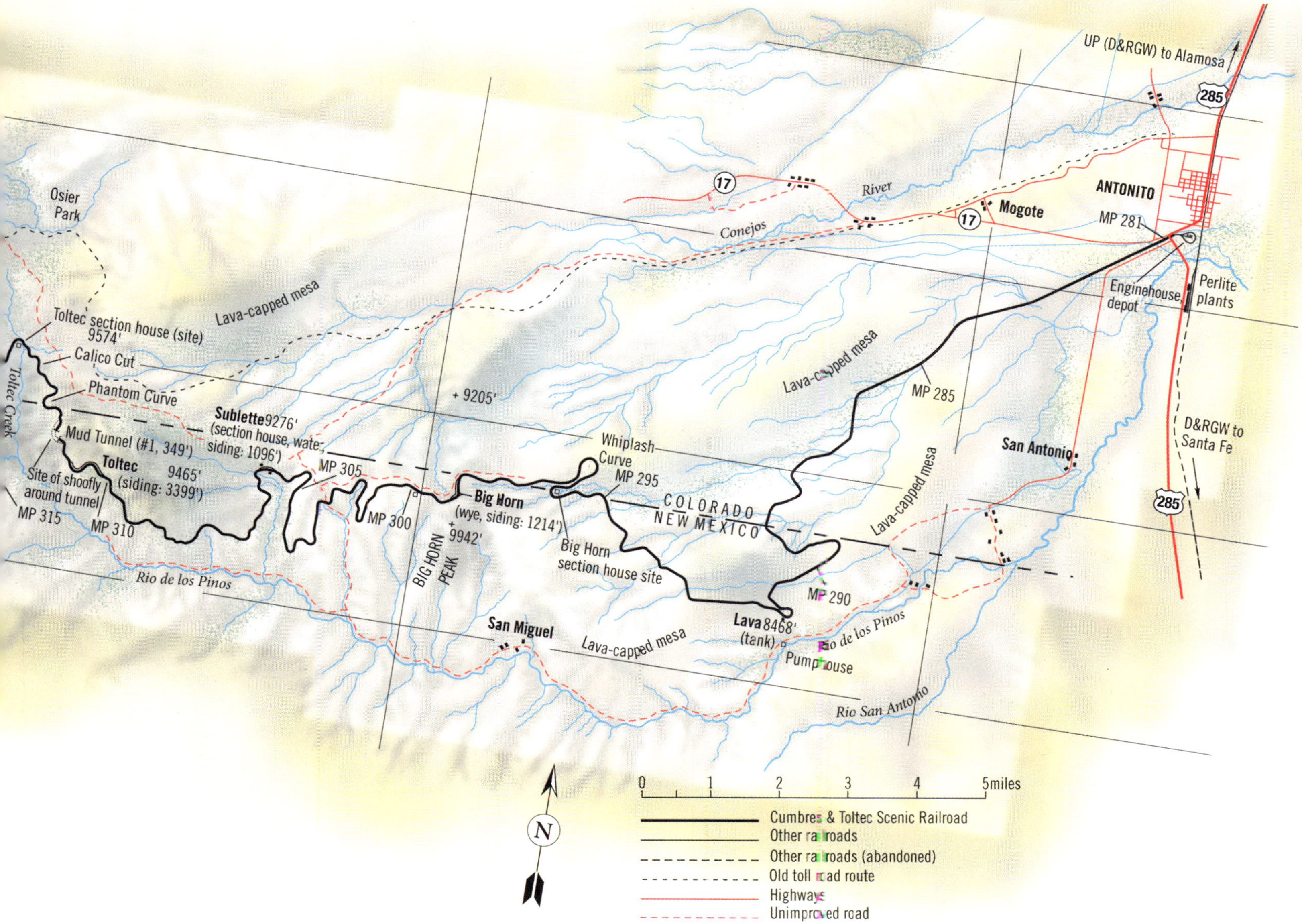

© 2001, Kalmbach Publishing Co., TRAINS; Robert Wegner

Cumbres & Toltec

Cumbres & Toltec

A photographic tribute to America's most spectacular scenic railway

By Sam Furukawa

P.O. Box 1073
San Carlos, CA 94070-1073

Edited and with text and captions by Bob Hayden

DUST JACKET: In a once-in-a-lifetime lineup, representatives of all three classes of Cumbres & Toltec steam locomotives smoke for the photographer at Cumbres pass, Colorado. From left to right, K-27 No. 463 (with white extra flags), K-37 No. 497, and K-36 No. 489. September 2001.

DUST JACKET BACK: Cascade Trestle is one of many highlights on the railroad's memorable trip through Toltec Gorge. Here, No. 484 leads a westbound C&TS train across the bridge on a brilliant autumn afternoon. September 2006.

PAGE 2: On a bitterly cold late-October day, C&TS K-27 No. 463 and K-36 No. 497 lean into the four percent grade to lift a special freight run from Chama to Cumbres. October 2002.

ISBN: 978-0-9790113-0-6
Printed in China by Everbest Printing Company Ltd., through Four Colour Imports Ltd., Louisville, Kentucky. First Printing, March 2007.

To order additional copies of this book, write to Book Publishing Division, NGPF, 7 South Hijo de Dios, Santa Fe, New Mexico, 87509-9133, or call 505-466-4970 between 9 a.m. and 5 p.m. Mountain Time.

Contents

The remnants of winter snowdrifts abound as C&TS No. 484 leads the first eastbound train of the season downgrade from Cumbres Pass toward Los Pinos and Osier. Melting snow has turned trackside ditches into temporary ponds and lakes. The calendar says it's been spring for more than two months! May 2005.

Meet Sam Furukawa

SAM FURUKAWA has photographed narrow gauge railroads all over the world, but the one he keeps coming back to is the Cumbres & Toltec.

His love of trains began with a Lionel O gauge set at age 3, and grew throughout his school years. As a high school student, he photographed the final years of Japanese steam.

Sam has been instrumental in establishing the Narrow Gauge Preservation Foundation, and in funding its programs (page 174). Now retired from a long career at the Microsoft Corporation, he devotes part of his time to recording the operations of the surviving narrow gauge lines, particularly those in Colorado and New Mexico, and pursues modeling in On3 and Sn3 scales. He also serves as Chairman of JMA (Japan Association of Model Railroaders).

When you ride the Cumbres & Toltec, it seems as if everyone – on and off the train – has a camera. Keep a sharp lookout – one of those photographers could be Sam!

Dedication

This book is dedicated to the railroaders who operate the Cumbres & Toltec, past, present, and future.

As a teenager Sam photographed the last days of steam in Japan. Here, a Class C62 4-6-4 leads a passenger run at Onuma Park on JNR's Hakodate line. Hokkaido Island's Mt. Komagatake looms in the background. March 1971.

CUMBRES & TOLTEC SCENIC R.R.
LOBATO
Cumbres & Toltec
487

Introduction

ALMOST HIDDEN FROM PUBLIC VIEW in a high and remote corner of the Rockies is the Cumbres & Toltec Scenic Railroad. The 64-mile line crosses and recrosses the border between southwestern Colorado and northern New Mexico, and passes through some of the most spectacular mountain scenery anywhere in the world.

If that weren't distinction enough, trains on the C&TS are powered by 19th- and 20th-century technology: steam locomotives. Those locomotives, the trains they pull, and the scenery they run through are the subjects of the photographs in this book.

The pictures speak for themselves, as do the trains and the scenery they depict. The history of the line, originally built by the Denver & Rio Grande in 1880 and 1881, has been covered in dozens of books already published, and is only mentioned here to explain what it is you're seeing.

Instead of more history, this book seeks to show what the C&TS looks like today. While you'll find plenty of photos of locomotives and trains, you'll also find pictures made to evoke the vast scenery that surrounds them.

Put another way, if you are among the hundreds of thousands of visitors that have ridden Cumbres & Toltec trains, the photos in this book are what your experience looked like from trackside – and beyond!

PRECEDING PAGES: Leaves are changing colors and beginning to fall as C&TS No. 487 leads a late-season train east out of Chama. The train has just crossed the Rio Chama bridge and Highway 17, and is headed for the stiff climb to Cumbres. September 2006.

RIGHT: A rich plume of black coal smoke tells us that No. 484 is working hard on the four percent grade to Cumbres. The train is approaching the road crossing near Coxo, Colorado, on the home stretch toward the pass. May 2005.

The essential mountain-railroad character of the Cumbres & Toltec is revealed when the trains are seen from a distance. Here, at Los Pinos, Colorado, an eastbound C&TS train is dwarfed by the tree-covered slopes that form the flanks of the valley. May 2005.

Coal Smoke over Chama Yard

CHAMA, NEW MEXICO, is the western terminus of the C&TS, and in many ways is the operational heart of the railroad. Here, as in D&RGW days and D&RG days before that, the C&TS maintains its locomotive shops, general offices, and associated service facilities.

Here, also, is stored the bulk of the railroad's remarkable collection of 19th- and 20th-century narrow gauge railroad equipment. (Narrow gauge refers to the distance between the rails. "Standard gauge" is 4´ 8-1/2″; anything less is "narrow gauge." On the C&TS the distance is three feet.)

Chama is a railroad town, meaning that its importance was established when the railroad arrived in 1880. That importance continued throughout nine decades of D&RG and D&RGW operations. All trains stopped here, many to change crews and replace or add locomotives. The town was peopled by railroaders and by those who provided services for them.

From Chama the railroad built west through New Mexico and Colorado toward the booming mining districts around Durango and Silverton, Colorado. Mining collapsed in the 1890s due to an act of the U.S. Congress which discontinued a policy of buying silver for coinage. Then the line hauled timber, coal, and livestock. Chama was an important lumber town, and in the 1930s became a transshipment point for crude oil bound to a refinery in Alamosa, Colorado.

There's a lot to see at Chama, and because the railroad is owned by the States of Colorado and New Mexico, the property is open to foot traffic. Just be careful!

It's almost train time at the Chama depot, where a guitar-strumming minstrel entertains passengers while the train crew prepares for today's trip. September 2001.

OPPOSITE PAGE: C&TS No. 484 on the Chama ash pit, with the tall coaling tower to the right. The raised track just beyond the locomotive, above the concrete wall, is used to deliver coal to the tower. May 2005.

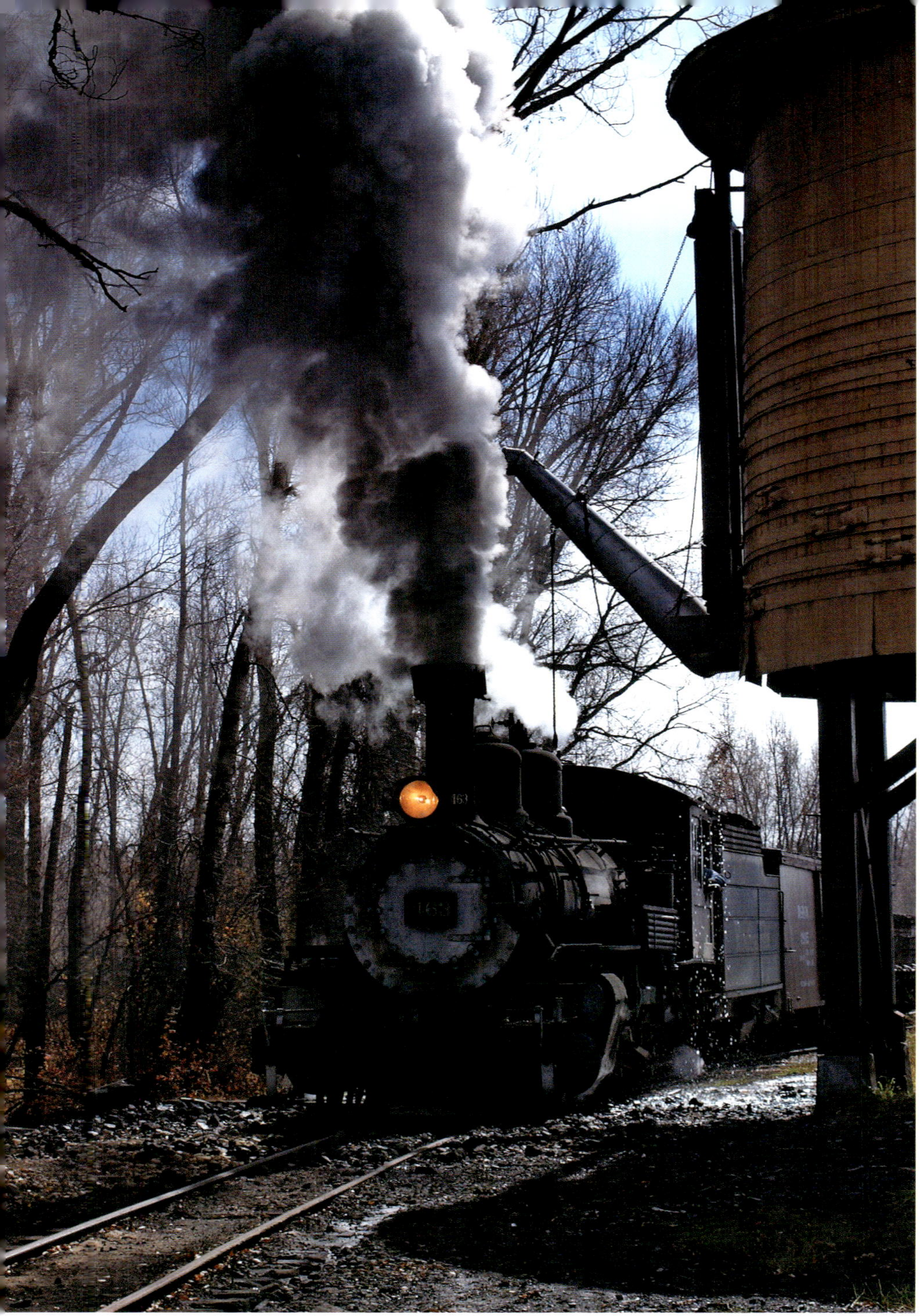

OPPOSITE PAGE: A view north through the Chama yard. The water tank and coaling tower are in the distance; the depot and coaling tower are closer, on the left. A special freight train is on the siding opposite the depot, and it looks like there's a snow-storm brewing in the high country today. October 2002.

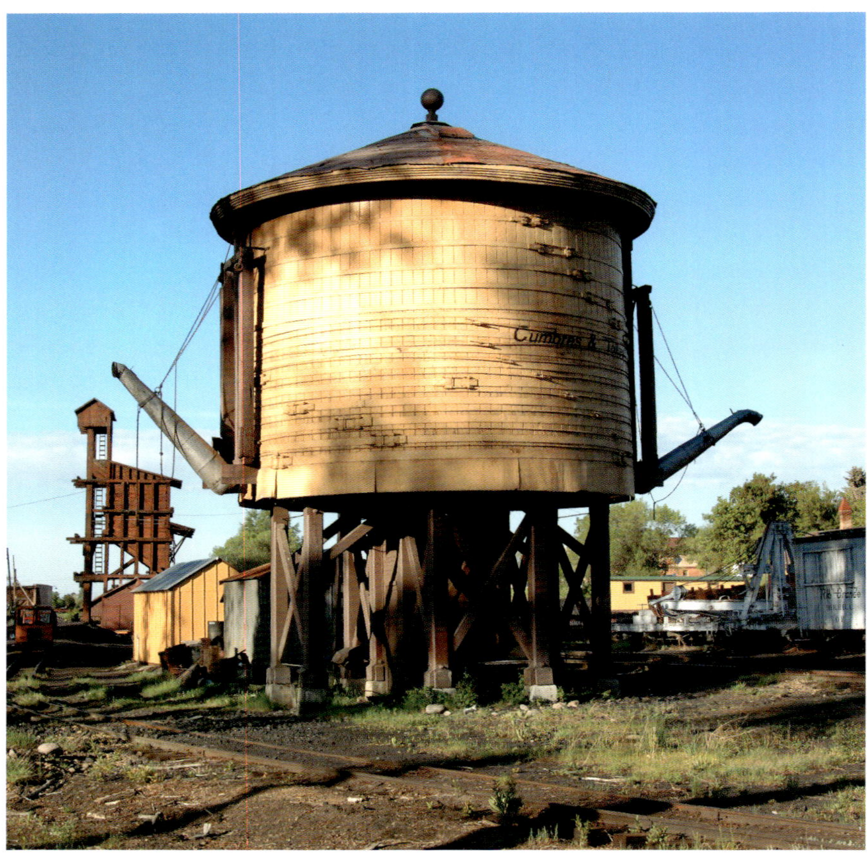

A classic Denver & Rio Grande Western design, Chama's water tank features dual spouts: one for the main line and another for the service track that runs over the ash pit. May 2005.

On a chilly autumn afternoon K-27 No. 463 assembles a freight train for a special, end-of-the-season event. October 2002.

Chama yard from on high! It's a few minutes before 10 a.m., and an eastbound train is set to depart. A second K-36 locomotive is in front of the shop (the red building). The town's main street (and State Highway 17) is on the embankment behind the tracks, and the C&TS parking lot is close to capacity. That means plenty of riders for today's trains! September 2006.

The extreme south end of the Chama yard is the location of the C&TS wye, used to turn the locomotives, and stockyard restoration. No. 463 has tied onto a stock car and is heading back to the depot with it. There's snow on the mountains ahead. October 2002.

Same locomotive, same task, different day. No. 463 is again hard at work making up a train, this time in the heart of the yard, near the depot. Newly painted (and as yet unlettered) box car No. 3605, at right, is a project of the Friends of the Cumbres & Toltec Scenic Railroad, a long-standing preservation group dedicated to maintaining the C&TS as a living museum of railroad history. September 2001.

Whistle screaming to warn motorists on Highway 17, No. 487 is on the wye at the south end of Chama yard. The triangular arrangement of track turns locomotives end-for-end for the trip back up the hill. May 2005.

OPPOSITE PAGE: C&TS No. 484 rocks into the track to the south end of the shops during early season switching chores at Chama. Two K-class engines are out of service and awaiting repairs at right. May 2005.

CHAMA
CHAMA
Cumbres & Toltec Scenic Railroad
WESTERN UNION
TICKET OFFICE

Buildings opposite the depot are devoted to locomotive maintenance. The two-stall brick structure on the left is the locomotive shop, where repairs are accomplished. The darker building to the right is the two-stall remnant of the Chama roundhouse, originally a much bigger facility. C&TS No. 488 is under steam in front of the shop, while No. 489 is disassembled and awaiting repairs. May 2005.

OPPOSITE PAGE: Today's train has departed, allowing a rare portrait of the Chama depot without a crowd. The structure is little changed from its D&RGW days. A National Park Service plaque (left) next to the door helps explain the C&TS to first-time visitors. June 2003.

Tall and seemingly ungainly, Chama's coaling tower is the most recognizable landmark in the yard. The highest portion of the structure, at the back, houses a bucket hoist mechanism that brings coal from below track level and dumps it into the coal pocket. From there, coal moves by gravity down to the chute, and when the gate is raised, into locomotive tenders. The tower has been out of service for several years, but plans are in hand to restore it to operation. May 2005.

Chama yard is alive with steam and smoke as C&TS No. 487 (right) and No. 497 back toward their train at the depot. It seems there are plenty of photographers on hand, too. May 2001.

A special train led by K-27 No. 463 and a cut of freight cars starts out of Chama and heads for Cumbres. Two more locomotives, Nos. 489 and 497, are farther back in the train, providing both additional tractive effort and additional drama. September 2001.

Oil became a significant freight commodity for the Cumbres pass line in the 1930s, when crude trans-shipped at this loading facilty at Chama began going to the Oriental refinery at Alamosa. The tank cars are maintained and used in C&TS freight specials. May 2002.

Here's more routine switching action on a sunny spring day in Chama. Engineer Jeff Stebbins, with his bowler hat, is backing No. 487 and caboose 0306 toward the north end of the yard, while the brakeman climbs aboard. May 2002.

The C&TS crosses the Rio Chama on this handsome two-span through truss bridge just north of town. While the D&RGW had similar bridges on other parts of its narrow gauge system, this is the only one on the C&TS. September 2006.

The Rio Chama bridge has been a favorite of photographers for over a century. The stately Ponderosa pine at left, one side scorched by 125 years of steam locomotive exhaust, is named the "Jukes Tree" for nineteeth-century railroad photographer Fred Jukes. (Left) May 2005, (above) May 2006.

A Railroad with Friends

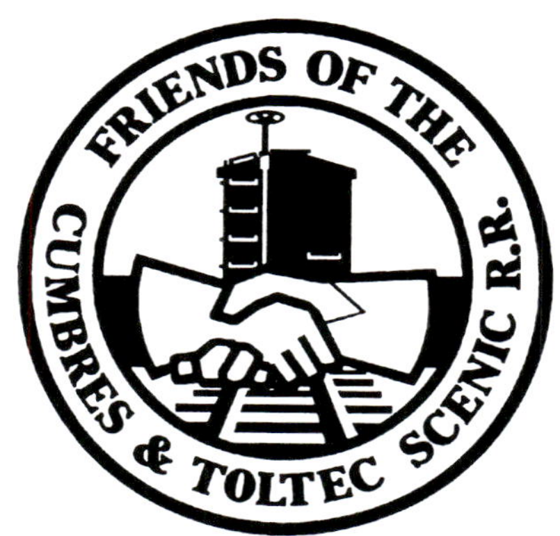

EVEN A RAILROAD can use some friends, and the C&TS has a lot of good ones. The Friends of the Cumbres & Toltec Scenic Railroad maintains its headquarters in Albuquerque, New Mexico, but you'll see a lot of their members – and a lot of what they've accomplished – in the yard at Chama.

Established only a few years after the states of Colorado and New Mexico purchased the railroad, the Friends has become a recognized leader in the railroad preservation field. With over 2,000 members worldwide, this non-profit organization is the official museum support group for the C&TS, managing the museum and restoration functions of the railroad.

As part of this role, the Friends has assumed the responsibility of preserving all of the railroad's historic assets that are not used in the regular tourist train operation. They also take a leading role in interpreting the railroad to the public.

In recognition of its efforts, the Friends has received awards from the New Mexico state legislature, the Colorado Historical Society, and the New Mexico State Historic Preservation Office.

In recent seasons the organization has increased its involvement by establishing non-profit subsidiaries for the purpose of operating the railroad's passenger trains.

To join the Friends, contact information is on page 31.

Each summer the Friends organize several week-long work sessions, and they get a lot done. Restoration of D&RGW drop-bottom gondola No. 811 (below), repainting a rotary snow plow (opposite page), and returning spreader OU to service (right) are typical projects undertaken by these skilled and dedicated volunteers. All photos June 2003.

The Brutal Climb to the Pass

THE GRADE FROM CHAMA to the top of 10,015-ft. Cumbres pass includes stretches of four percent, which means that the track rises four feet for every 100 feet traveled. Such a rise may not seem steep for your family car, but in railroading it poses a brutal challenge to machines and men.

The Denver & Rio Grande planned and built this 14-mile section of its San Juan Extension as a helper grade, a stretch where trains of normal length would require an additional locomotive, sometimes more than one. Chama was where the helper locomotives were stationed and maintained, and eastbound trains stopped there while the helpers were added.

This operational plan made for 125 years of spectacular railroading, which continues today. In the early days of small locomotives, three, four, or even five of them would be required to lift trains to Cumbres. When the larger K-class engines arrived, eastbound trains would call for two locomotives, or long trains would have to be taken to the pass in two or more pieces.

Most regular C&TS trains can surmount the grade with a single K-class locomotive, but extra passengers or freight specials call for doubleheading. The result is nothing short of earth shaking: Steam locomotives working at full throttle, belching coal smoke high into the air and echoing thunderous exhausts off the walls of the Wolf Creek valley.

Doubleheaded K-36 locomotives lead a long passenger train out of The Narrows and toward the aspen grove near Lobato trestle. The Narrows marks the beginning of the four percent grade, and the engines are working hard. August 2006.

OPPOSITE PAGE: Locomotives 463 and 497 lean into the grade toward Cumbres. The train is a special freight for photographers, with the engines relettered for the D&RGW. Staged after the end of the regular C&TS season, it's one cold event, and there's snow ahead in the high elevations. October 2002.

Topped off with coal, water, and sand for the strenuous trip up the hill, C&TS No. 487 backs toward the Chama depot and today's eastbound train. June 2003.

OPPOSITE PAGE: On a different morning almost two years later, No. 484 rolls past the Chama tank and through cottonwood shadows on the way to the Chama River crossing. May 2005.

Cumbres & Toltec

The whistlepost at the north end of Chama yard ("W," next to the tender) alerts the engineer to sound a warning as No. 484 and its train approach the Rio Chama bridge. May 2005.

OPPOSITE PAGE: On a perfect spring morning C&TS No. 487 crosses the river and passes the Jukes tree. A light breeze is moving just slightly faster than the train. May 2006.

OPPOSITE PAGE: The Jukes Tree again, this time from above! Fall colors are near their seasonal peak as No. 487 puts on a great show before heading for the high country. September 2006.

The line between the Highway 17 crossing north of Chama and The Narrows is straight and almost level. This double-headed freight is about to enter The Narrows and encounter the first stretch of four percent grade. October 2002.

Called "The Narrows" because the Rio Chama cut a steep, V-shaped valley here, the C&TS track is squeezed between the waterway below and the paved highway above. The Narrows got even narrower when parallel Highway 17 was widened in the 1970s, necessitating the long stretch of steel retaining wall shown here. September 2006.

Vibrant spring greens abound as No. 488 leaves The Narrows and heads for Lobato with an almost clear stack. The high rock wall above the locomotive is typical of the landscape along this stretch of the railroad. May 2005.

C&TS No. 487 turns the corner and heads for the aspen grove just south of Lobato, where Wolf Creek joins the Rio Chama. Deer are often sighted in the grove. The rock walls on the other side of the valley stand tall ahead of the train. There's a sharp curve ahead, and a 10 mph speed limit. June 2003.

There's coal smoke in the air as No. 484 arrives at Lobato and whistles for the bridge with one of the first trains of the season. The phony water tank here was built as a movie prop, as was a small depot that has since disappeared. The stock pens are the real thing, and were used almost until the end of D&RGW operations in the 1960s. May 2005.

Lobato's steel trestle dates to 1883, when D&RG locomotives were much smaller than the current K-class engines. Because of a weight restriction in place since early in the 20th century, doubleheaded trains must cut off the helper engine to cross the bridge, as No. 497 is doing here. The road engine, No. 487, and train then follow, and the helper couples up on the north side of the trestle. Both photos May 2002.

We glimpse No. 488's fire as another shovel of coal is fed into the inferno. The train is about to cross Lobato trestle, headed for a tight 10-mph curve in the cut beyond. The 100-foot tall bridge is barely visible from track level. May 2005.

ABOVE: Almost obscured by her own smoke and steam, No. 463 pounds upgrade on a photo freight east of Lobo Lodge and Dalton. October 2002.

LEFT: Exhaust, whistle, and a boiler blowdown make for action at Lobato. September 2006.

45

Cumbres & Toltec
487
Rio Grande
497

ABOVE: Same location, less smoke. No. 488 is only a minute or two from crossing into Colorado for the first time. Most of these cars have been repainted in the new C&TS red. May 2005.

LEFT: The four percent grade is still taking its toll, and both No. 484 and her fireman are working hard. The train is exiting the reverse curve and is continuing toward the state line. May 2005.

OPPOSITE PAGE: It's a sunny day, but coal smoke casts a pall over C&TS 497 and 487 and their train. The location is east of milepost 337, below Highway 17. May 2002.

Beneath much smoke and steam, another doubleheader gets underway from Cresco, Colorado, this time on a special freight run. With a train of five tank cars, eight stock cars, and an extra caboose, the second locomotive is not just for show. October 2002.

OPPOSITE PAGE: Cresco Tank is a regular stop for trains headed upgrade to Cumbres. This doubleheaded run includes a water car behind the locomotives to wet down the right of way due to fire danger. The slender white cross is a memorial to Friends of the C&TS member Mike Hipskind, who collapsed and died near the spot several years ago. May 2002.

PRECEDING PAGES: C&TS No. 487 whistles for the road crossing near Coxo, Colorado, as traffic on Highway 17 pauses to take in the spectacle of steam mountain railroading the 21st century. August 2006.

Another special photo train climbs the last stretch of four percent at Coxo, headed to Windy Point and Cumbres. For authenticity, passengers ride in some of the original C&TS converted boxcars first used in the 1970s. August 2006.

The rugged and colorful volcanic formations at Windy Point have witnessed the passage of narrow gauge trains for more than 125 years. The track here is on a narrow ledge, and most photos must be shot from far below. Both these views show special doubleheaded runs that evoke D&RG railroading of a century before. (Left) August 2006. (Above) October 2002.

This short timber trestle carries C&TS trains over the old highway cut and into the yard at Cumbres. No. 487 is still working hard to surmount the last few feet of four percent grade, but she'll find easier going on the way down to Antonito, almost 50 miles away. September 2001.

RIGHT: The pass at last! Eastbound C&TS trains pause for 10 minutes at Cumbres, while the locomotive takes water and the crew inspects the train. The D&RGW's Cumbres depot is long since torn down, and the section house, far right, serves to mark the spot. September 2001.

OPPOSITE PAGE: The fireman can rest, at least briefly, as No. 488 crests the Cumbres summit at 10,015 feet above sea level. Behind the train is Windy Point, and beyond that the beautiful Wolf Creek valley. May 2005.

First order of business for trains arriving at Cumbres is to refill the locomotive tender with water. No. 488, freshly painted and lettered, takes a big drink on an eastbound run. The standpipe is fed from a cistern in the hillside to the left, behind the photographer. While the fireman handles that chore, the engineer will inspect the running gear and oil around. May 2005.

Brief Pause at Cumbres

AS THE EAST END of the C&TS helper district, Cumbres is where helper engines cut off from their trains, turn, and return downgrade to Chama.

The pass is 10,015 feet above sea level, and snowfalls of up to 35 feet for a single season have been recorded there. Snow is often in evidence at both the beginning and end of the C&TS season, and it's not unusual to encounter blizzard conditions at Cumbres in September.

Before the introduction of powerful steam-powered rotary snowplows in the 1890s, much of the track at Cumbres, including a turntable, was covered with wood snowsheds as means of keeping the line open during the winter.

Under the current C&TS timetable, all trains make lengthy stops at Cumbres. Locomotives take on water, but the main reason for the stop is safety: All trains leaving Cumbres, regardless of direction, are headed downgrade, and their brakes must be in full working order before beginning the descent.

Just to the east of Cumbres is Tanglefoot Curve, a dramatic balloon loop where the line turns more than 180 degrees to make an elevation change of almost 40 feet. For those riding the trains, it's one of dozens of memorable vistas.

No. 484 has taken on water and the crew has completed its inspection of the train during its 10-minute stop at Cumbres. Now locomotive and train have crossed Highway 17 for the last time and are on their way eastbound to Osier, for lunch. May 2005.

This morning's eastbound train required two locomotives for the climb to Cumbres. The helper, No. 487, has cut off from the train and moved onto the track to the wye, while No. 497 takes the train on toward Osier, Toltec Gorge, and Antonito. The helper will now turn on the wye, take water, and return light to Chama. May 2002.

Today's helper is K-27 No. 463, and here she is being turned on the wye at Cumbres. The first step (above) is for Engineer John Coker to back onto the tail track, under the short remnant of snowshed. The brakeman throws the switch (left) and No. 463 runs forward to complete the move. Both photos, October 2002.

With her turnaround move complete, No. 463 exits the wye and backs onto caboose 0503 on the main line. Once coupled up, the helper engine is ready to run light back to Chama and assist on the next "hill turn" – a trip to bring another section of the train up the grade to the pass. Both photos, October 2002.

Safely past the Highway 17 crossing with her train, No. 488 shows only a wisp of exhaust steam from her turbogenerator. It's mostly downhill all the way to Antonito, and the fireman will have a lot less to do than on the battle upgrade. The track to the left is a passing siding where cuts of freight cars were stored in D&RGW days while long freight trains were assembled for the trip to Antonito. May 2006.

Eastbound trains leaving Cumbres begin their descent by negotiating Tanglefoot Curve, a dramatic balloon loop that drops the track almost 40 feet. Needless to say, Tanglefoot has been a favorite of photographers ever since they began coming to the railroad, over a century ago. These four photos, all taken from the same vantage point, show a train entering the curve (left), near its midpoint (below), returning (to right), and exiting the curve (bottom right). May 2005.

Eastbound toward Antonito

A few minutes ago this train came off the loop at Tanglefoot Curve and is now heading gently downgrade toward Los Pinos, Colorado, following Cumbres Creek. Highway 17, where the photographer is standing, has descended from its summit at Cumbres pass, and is also headed for Los Pinos. May 2005.

Not far from the location on the opposite page, this shot of No. 487 shows how green the high country can become by the middle of June. In addition to the spring greens of grasses, yellow wildflowers have begun to appear. June 2003.

AS EASTBOUND C&TS TRAINS DESCEND into the valley of the Rio de los Pinos and the Toltec Gorge, the scenery becomes more and more breathtaking. At Los Pinos the paved highway departs from the railroad line and heads over La Manga Pass, leaving the trains in high and isolated grandeur.

The trip through the gorge is quite different from the laborious ascent from Chama to Cumbres. Loud staccato exhausts and great volumes of smoke are replaced by soft chuffs and light exhaust. In many stretches the wheels furnish most of the train noise.

From Cumbres to Antonito is 50 miles, and the trains pass through some of the most varied and interesting scenery in the Intermountain West. Highlights include the deep valley of the Rio de los Pinos, tall Cascade trestle, the deepening Toltec Gorge, Rock and Mud tunnels, Phantom Curve, Whiplash Curve, and the long run through sagebrush country into Antonito.

Evidence of human settlement is scarce, with a few vacation homes at Los Pinos and small ranches in the valley below. Trains reaching Osier, elevation 9,637 ft., stop while the passengers have lunch in the large catering facility there.

Perhaps most significant is that the line is substantially unchanged from its construction in 1880. Here is a 19th-century railroad (albeit with 20th-century locomotives and trains) alive and well in the 21st century.

ABOVE: No. 487 has a clear stack, and the only exhaust rises lazily from her turbogenerator as she slowly drifts downgrade toward Los Pinos. The locomotive's coal supply, still piled high, will last all the way to Antonito. August 2006.

LEFT: Whistle screaming a warning note, No. 487 rounds a sharp curve before heading down into the Rio de los Pinos valley. May 2006.

OPPOSITE PAGE: Clouds appear to be gathering for an afternoon thunderstorm as No. 487 leads an eastbound train away from Cumbres. The yellow spot in the distance is the patrol speeder that follows the train to extinguish lineside fires. Its operator also occasionally retrieves hats lost by riders in the open car. August 2006.

OPPOSITE PAGE: A special freight with pipe gondola, idler cars, and six original C&TS converted boxcars carrying passengers crosses the trestle over the Rio de los Pinos, deep in the valley. The stream is only a trickle now, but much of the valley floods during the spring snow melt. August 2006.

BELOW: Well past the tank, a different train turns 180 degrees at its furthest reach into the valley, then runs back out. This is where Highway 17, the paved road above the train, parts company with the railroad. Both are headed for Antonito, but the road goes up and over 10,230 ft. La Manga pass, while the C&TS follows an easier alignment along the Rio de los Pinos. May 2005.

ABOVE: Turning sharply north, eastbound C&TS trains roll into the deep valley and are greeted by Los Pinos tank. A few vacation homes dot the valley, which is often inaccessible during winter. May 2005.

An unusually long C&TS train – 12 cars – rolls toward Osier along the east wall of the Los Pinos valley. Los Pinos is at an elevation of 9,706 ft., and the ridge line above the train exceeds 10,500. August 2006.

LEFT: This 137-foot-high steel trestle carries the C&TS over Cascade creek. Built in 1889, it is an exceptionally graceful design. Unlike the straight bridge at Lobato, Cascade trestle is curved. September 2006.

BELOW: Next stop: Osier – and lunch! No. 488 rounds yet another curve on the way to the most important stop between Cumbres and Antonito. The Rio de los Pinos valley, off to the left here, is deepening as we approach Toltec Gorge. August 2006.

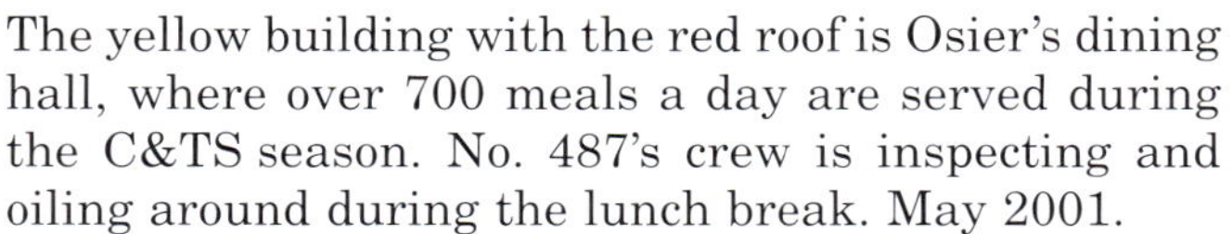

ABOVE: Osier, Colorado, is where the trains meet and the passengers eat. The standard operating scheme calls for the locomotive from Antonito to couple onto the train that came from Chama, and take it back there. The engine from Chama does the opposite, going on to Antonito. On this day the train from Chama is a special photo freight, which requires extra switching moves to return there. August 2006.

The yellow building with the red roof is Osier's dining hall, where over 700 meals a day are served during the C&TS season. No. 487's crew is inspecting and oiling around during the lunch break. May 2001.

A thousand photographs could not adequately convey the vastness and loneliness of the scenery around Osier, but this shot of a departing eastbound train comes close. The deep gash beyond the train is the ever-deepening canyon of the Rio de los Pinos, which will form 900-foot Toltec Gorge a few miles to the east. September 2006.

For the most part, the gentle descending grade to Antonito closely follows the contour lines of the topography. This part of the line east of Osier does just that, curving around the hillside to remain at nearly the same elevation. While this engi-neering strategy eliminated the need for extensive cutting and filling and allowed the D&RG's San Juan Extension to be built quickly, it meant that trains would forever have to run slowly over the sharp twists and turns. September 2006.

Rock Tunnel was one of only two bores on the D&RGW narrow gauge system; both are on the 64-mile C&TS line. This is its west portal, cut, like the rest of the tunnel, through solid rock. The cut stone retaining wall that leads to the portal makes this one of the most photogenic (though hard to reach) locations on the C&TS.

A few yards west of the retaining wall is this monument to President James A. Garfield, who was assassinated in 1881. Behind it, Toltec Gorge drops 800 feet to the stream bed. Both photos, September 2006.

The fantastic volcanic spires at Phantom Curve dwarf No. 484 and an eastbound train. A tall aspen at the height of its autumn color provides a counterpoint. September 2006.

Less than a mile farther east, Mud Tunnel was cut through volcanic deposits similar to those that produced the pillars at Phantom Curve. The soil's lack of stability required timber tunnel portals and a full lining to keep dirt and rock off the track. September 2006.

BELOW: Mud Tunnel from the inside, looking out. The little yellow building dead ahead is a railroad phone booth, where train crews could call in to the dispatcher to report problems or progress. September 2006.

Sublette, New Mexico, is yet another remote and lonely place on the C&TS. The historic buildings here have been restored by the Friends of the C&TS, and include a log bunk house (ahead) and the section house (right). September 2006.

OPPOSITE PAGE: It's a sparkling autumn afternoon as No. 484 passes the west end of Toltec Siding, elevation 9,465 ft. As the train moves east, aspens are more and more in evidence. September 2006.

Whiplash Curve, between mileposts 298 and 295, doubles back on itself twice to drop the line more than 200 feet. As the C&TS descends toward Antonito the vegetation changes dramatically. Evergreens and aspens give way to Ponderosa pine, pinon, and junipers, then chamisa and sage. The next 14 miles are a slow drift downgrade through open sage country. Both photos, September 2006.

Antonito, Colorado, elevation 7,888 ft., is the eastern end of the C&TS. Facilities here include the depot (far left), water tank with shop behind, a coaling platform, and a balloon loop for turning locomotives and trains.

It's late afternoon, and the Antonito train is coming in with its load of tired – but happy – riders. Both photos, September 2006.

Locomotives: Care and Feeding

All three classes of C&TS steam locomotives line up, belching coal smoke, for a memorable photo at Cumbres. From left to right: K-27 no. 463 (with white extra flags), K-37 No. 497 (with snow plow), and K-36 No. 489. September 2001.

AMONG THE MANY CHALLENGES of operating the
C&TS is maintaining its fleet of historic steam locomotives.
By almost any calculation all of these large and complex
machines are antiques: The newest entered service with the
Denver & Rio Grande Western in 1930, and has completed
three quarters of a century of service. The oldest was built
more than a century ago.

There are three classes of ex-D&RGW 2-8-2 locomotives on
the C&TS roster. The "K" stands for the Mikado, or 2-8-2,
wheel arrangement, and the number refers to the tractive
effort each locomotive can produce.

K-27 The sole example of this class is No. 463, built in 1903.

K-36 All of the 480-class locomotives are K-36s, built in 1925.

K-37 These 490-class locomotives were built by the D&RGW's
Burnham Shops in Denver in 1928 and 1930, using boilers
from retired standard gauge 2-8-0s originally built in 1902.

Every bit as important, though less often photographed,
are the logistic and maintenance systems that support the
locomotives. Coal, water, sand, and lubricating oil are
required on a daily basis, and all but water must be brought
from a distance. Repairs to these ancient machines must be
made on both as-needed and scheduled bases.

Perhaps most crucial of all – and all but invisible – is the
in-depth practical knowledge of steam locomotive operation
possessed by C&TS crews and shop personnel. Seventy-five
years ago, when steam was still king on all the world's railroads, such knowlege was widespread; now it is expertise
held by a dwindling cohort – a lot of its members to be found
at Chama, New Mexico, and Antonito, Colorado.

No. 487 is a K-36 class locomotive, built in 1925.
The coaling tower at Chama is a critical part of the
support facilities for steam, as are the sand house
and sand storage bin to the right. May 2002.

When built, hulking No. 497 was among the largest narrow gauge locomotives anywhere in the world. The K-37s were produced by mating boilers from old standard gauge 2-8-0s to new narrow gauge 2-8-2 running gear. The result was – and is – a success. May 2001.

K-36 No. 484 poses for a portrait at Cumbres, where a spring snowdrift reflects light onto her running gear. These locomotives of the 480 class currently make up the mainstay of C&TS motive power. May 2005.

Unique on the C&TS locomotive roster is K-27 class 2-8-2 No. 463. Constructed by Baldwin Locomotive Works of Philadelphia, Pennsylvania, in 1903 and much modified since, this is the oldest and smallest steam engine on the C&TS. It is owned by the City of Antonito, and as of this writing it is out of service and awaiting repairs. May 2001.

Even the best grades of Colorado and New Mexico coal don't burn completely, so each C&TS locomotive must visit the ash pit at least daily so cinders and clinkers can be cleaned from its ash pan. No. 484 is on the pit here, and the flames testify to the inferno in her firebox. May 2005.

K-36 class 2-8-2 No. 488 has more than 81 years of service under her belt as she waits to lead a train west from Antonito. The track in the foreground appears to be ballasted with cinders from the ash pit. The small boxy structure on the tender, behind the coal pile, is the "doghouse," a shelter where the head-end brakeman rode on D&RGW freight trains. September 2006.

That's enough! In fact, it's more than enough, and No. 463's tank is full of water. This standpipe is at Cumbres, and is fed by a cistern in the hillside above. The path leading off to the right goes up the hill to the cistern and its capacity gauge. September 2001.

K-27 No. 463 in an unusual tender-first pose on the ash pit track behind the Chama coaling tower. The smallest locomotive on the C&TS steam roster, No. 463 entered service with the D&RG in 1903. May 2001.

No. 487's morning routine includes a stop at the coal stage, where a backhoe fills the bunker in the tender.

The rounded dome immediately behind No. 487's stack holds sand, which can be piped to the rails to improve traction. Here, one of the engine crew is emptying bags of dry sand into the dome for the trip up the hill. Both photos, June 2003.

The final member of the K-36 class, No. 489, simmers quietly in front of her eastbound train at the Chama depot. May 2001.

Lubrication is vital to reduce wear and to keep the venerable K-class locomotives running smoothly. (Above) Shopman Jose Torres forces grease into a crosshead bearing with a pressure gun. (Left) Grease protruding from the rotating joints of an eccentric crank proves that the job has been done thoroughly. Both photos, May 2005.

There's no turning off a steam locomotive overnight – once that big boiler is heated up, it has to be kept hot in order to be ready for service in the morning. This time exposure shows No. 487 outside the shop at Chama, with fire banked, turbogenerator running, and steam up for tomorrow's train. June 2003.

Coal-burning steam locomotives operate in a filthy and abrasive environment, so washing off the grime is part of the daily routine. Engineer Soni Honigger has broken out the steam hose to clean up No. 463 (left) and No. 484 (below). A clean engine is a happy engine! Left, September 2001. Below, May 2005

A check of the air ringer on 487's bell comes next, with a wrench at the ready. May 2006.

Jeremy Garcia carefully cleans the headlight glass on 487, while the next generation of C&TS engineer – maybe – looks on. September 2001.

OPPOSITE PAGE: The daily routine continues, as classification lamps are hung on No. 487's smokebox. May 2006.

OPPOSITE PAGE: Osier is a great place for a clear view of No. 487's running gear. The large appliance near the center of the photo is a steam-driven air compressor, which provides critical power for braking. August 2006.

There's considerable irony in calling steam locomotives "iron horses," because they require almost as much care as flesh-and-blood horses. These are machines from an earlier age, when dozens of men could be devoted to keeping them running. Here, No. 484 is in the shop at Chama for minor adjustments to her running gear before the 2006 operating season. Shopman Tommy Garcia is probably on his way to get a bigger wrench. May 2006.

K-36 No. 489, with cab removed and firebox marked for high-tech testing, will require many hours of attention from the shop force before returning to service. May 2005.

98

Railroading is hard work, but that doesn't mean nobody smiles. Ronnie Lopez (left) and Marvin Casias look down from the cab of K-37 No. 497 in a moment when the main task seems to be waiting. May 2002.

Men at Work: C&TS Railroaders

THE C&TS EMPLOYS MORE THAN 60 people during its May-to-October operating season, and a handful work on a year-round basis. Many of them are behind the scenes, taking reservations, preparing the payroll, and handling the abundant paperwork that has always characterized a railroad.

The operating railroaders – train crews – are the people that go with you on your trips over the C&TS. They run the locomotives, switch the cars, and help passengers on and off the trains.

You'll occasionally get a glimpse of shop personnel, particularly at Chama and Antonito. These skilled individuals are critical to keeping equipment, both old and new, running.

Every bit as important are the maintenance of way crews that care for the 64 miles of track and other aspects of the railroad's physical plant. Largely unseen by those riding the trains, these crews are vital to smooth – and safe – trips.

This short chapter cannot hope to depict all the C&TS people; instead, it's a representative sampling of the railroaders at work between Chama and Antonito at any given time.

Engineers Earl Knoob (on the ground) and Jeff Stebbins shake hands after a double-headed trip up the hill to Cumbres. Fireman Nathaniel Miller looks on. September 2001.

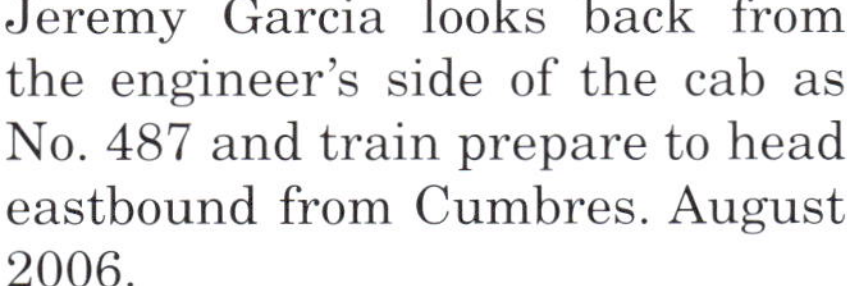

Jeremy Garcia looks back from the engineer's side of the cab as No. 487 and train prepare to head eastbound from Cumbres. August 2006.

Longtime Chama Shop Supervisor Steve Montano takes a brief break to smile for the camera during preparations for the start of the 2005 C&TS season. There's still lots to do. May 2005.

Conductor Allan Loomis confers with Engineer Jeff Stebbins during the stop at Cumbres. The gray device in Stebbins' hand is an infrared heat monitor, one of the few concessions to high technology in 21st century C&TS operations. May 2005.

At the throttle of No. 487, Engineer Earl Knoob looks ahead for hand signals while a helper is cut off at Cumbres. May 2002.

Conductor Frank Stapleton motions No. 463 back through the yard at Chama during a routine switching move. September 2001.

During switching moves at Chama, Brakeman Mike Lawrence signals the engine crew. May 2002.

Nathaniel Miller rides the tender of No. 497 as the locomotive backs through the yard at Chama and toward her train. May 2001.

Conductor Ray Martinez in an animated discussion of what to do next. While brakemen and conductors may wear identical garb, the conductor is the boss of the train. May 2006.

Steam railroading is an arduous calling, and it makes men tired. Michael Rivas in a contemplative pose with No. 489 outside the roundhouse at Chama. May 2001.

There's also a trace of weariness in Engineer Soni Honigger's gaze as he looks back from the cab at the end of a run at Chama. May 2005.

Speeder patrols – one ahead of and one behind each train – are an essential part of C&TS operations. In radio contact with the trains and with the C&TS dispatcher, the crews of these gasoline-powered vehicles monitor track conditions and check for lineside fires sparked by the locomotives. Dave Sands is on the footboard of speeder 012, at Cumbres, which hauls a tank trailer for fire fighting. May 2001.

Brakeman Don Brown and veteran shopman Tommy Garcia confer near the depot as a special train is assembled at Chama. September 2001.

Shop Foreman Ed Beard takes a turn at the throttle of No. 497 to ensure the locomotive is ready for the day's work. September 2001.

Conductor Allan Loomis discusses today's the trip up the hill with Fireman Tracy Carroway while No. 487 takes water at Cumbres. June 2003.

One of the veteran trainmen on the C&TS, Carlos Llamas is firing today. He's smiling, too. May 2005.

Something – or someone – has called for smiles as engine crew Soni Honigger (left) and Jeff Stebbins look back over their train. October 2002.

Conductor Sam Roybal wears a portable radio on his hip. Another concession to modern technology, these radios are a key tool for safety in C&TS operations. May 2005.

The work of the shop force never ceases. Tommy Garcia pumps grease into a fitting on a K-class locomotive crosshead. May 2006.

Steve Montano attaches flags to the smokebox brackets of K-36 No. 487 to help celebrate Flag Day. June 2003.

After a long career in railroading elsewhere, Frank Turner served as president of the C&TS Management Corp. for the 2006 operating season. In this photo he's just completed a long day as the motorman of speeder MW-02. September 2006.

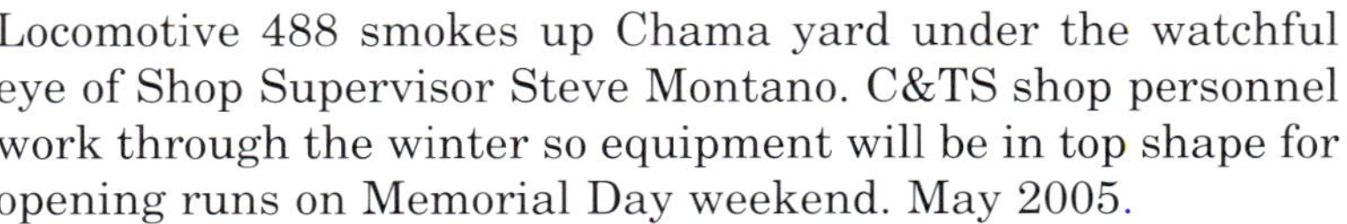

Locomotive 488 smokes up Chama yard under the watchful eye of Shop Supervisor Steve Montano. C&TS shop personnel work through the winter so equipment will be in top shape for opening runs on Memorial Day weekend. May 2005.

Grime is ever present in railroading, and steam locomotives produce more of it by the minute. Hostler Juan Torres wears his share of the stuff as he goes about his duties on No. 497 in the Chama yard. May 2002.

Engineer Earl Knoob lets us know that everything is OK with a big smile and an affirmative wave. May 2002.

Fireman Tracy Carroway climbs into the gangway of K-37 No. 497 after taking water at Cumbres. Next stop: Lunch! May 2002.

Though not on the C&TS payroll, Ralphie was a regular member of the crew at Chama during the 2002 season. A white companion yard dog, Aspen, wouldn't pose for a photo. October 2002.

Westbound to Osier

OPPOSITE PAGE: Snow has arrived early in the higher elevations of the towering Sangre de Cristo mountains, and the 14,000-foot peaks form a dramatic backdrop for a C&TS train departing Antonito. September 2006.

D&RG locating engineers surveyed the line from Antonito to Cumbres with maximum grades of 1.42 percent, reasoning that most freight traffic would go in that direction. After gold and silver mining collapsed in the San Juan, most freight moved east. K-36 No. 488 is the standby engine as No. 484 prepares to head west from Antonito with today's train. September 2006.

Wreathed in steam and smoke, No. 484 seems to be emerging from a cloud of her own making as she starts the train out of Antonito at 10:00 on a chilly autumn morning. The Antonito engine house is behind the engine, and the depot is out of the photo to the left. September 2006.

THE SAN JUAN EXTENSION of the Denver & Rio Grande built west from Alamosa, Colorado, in 1880. Its goal was to reach the rich mining districts that eventually became Durango and Silverton, Colorado.

The new railroad ran a straight and relatively level line from Alamosa to Antonito, which was established as a railroad town and a junction where another new line headed south to Santa Fe, New Mexico. Then the line to Durango turned to the west to assault the San Juan mountains.

D&RG locating engineers expected most of the bulk traffic on the line would be headed west, with only concentrated ores and precious metals going east. For that reason, they planned the westbound line with less than a 1.5 percent grade, and put the 4 percent climb on the eastbound leg.

But the rules changed when the U.S. Congress repealed the Silver Purchase Act in 1893, effectively sweeping away the economic premise on which the San Juan Extension had been built. Instead of gold and silver, lumber, coal, and livestock furnished the traffic, most of it headed east. A flurry of activity due to the oil and gas boom around Farmington, New Mexico, kept the line viable into the 1960s.

Westbound C&TS trains lift themselves at a steady 1.42 percent (a rise of 1.42 feet for every 100 feet travelled) all the way from Antonito to Cumbres. Passengers on these trains experience the gentle climb through the sagebrush country, then into lava-capped mesas and other volcanic landscapes.

Once past the stop at Sublette, New Mexico, the railroad begins to follow the course of the Rio de los Pinos, which includes the dramatic Toltec Gorge.

This is high, desolate, and lonely country, and the human presence is limited. When the railroad built through 125 years ago it brought new population, but most of it was railroad operating personnel. Even today, once you leave Antonito you'll probably see more cows than people, if you don't count your fellow passengers!

The long straightaway just west of the Highway 285 road crossing in Antonito is a good place to sight along the train to look for trouble The conductor on the rear platform is doing just that, as No. 484 brings the train up to speed for the run through the sage country and into the mesas and Toltec Gorge. It's going to be a beautiful trip. September 2006.

This westbound train is briefly headed east toward the snowy Sangre de Cristo mountains as No. 488 enters Lava Loop. The track turns more than a half circle here to gain elevation while keeping the grade at less than 1.5 percent. A cutoff between the two sides of the loop, barely visible in the sagebrush at right, was used to turn rotary snow plow trains during the years of D&RGW operations. September 2006.

West of Lava Loop No. 484 heads upgrade while traversing the middle level of Whiplash Curve. This three-tiered arrangement of loops lifts the C&TS out of the sagebrush country and into the mesas toward Big Horn. The top tier is visible as a shelf crossing from right to left above the train. September 2006.

No. 488 rolls upgrade under an umbrella of her own coal smoke and across the short trestle on the lower 180-degree curve at Whiplash. The D&RGW's Big Horn section house was located here, at Milepost 296, many years ago. September 2006.

Our final view of Whiplash shows all three levels, but you've got to look closely. The lowest tier enters the photo at upper left, and makes a 180-degree turn just below the bright yel-low aspen groves above the caboose. The gas speeder above No. 484's headlight shows the location of the upper tier. Big Horn Peak rises in the center of the picture. September 2006.

When we reach Big Horn, at 9,022 ft., most of the trees are Ponderosa pine and aspens, and much of the sagebrush has been replaced by chamisa. The wye here is to the right of the train, and the passing siding to the left was where trains often met in D&RGW days. September 2006.

Shot from the flank of Big Horn mountain, C&TS No. 484 and train are just west of Big Horn, continuing the gentle climb toward Sublette, Osier, and Cumbres. September 2006.

Sublette is deep in Canada Jarosita, a side draw formed by a tributary of the Rio de los Pinos. This westbound train, C&TS No. 215, is slowing for its scheduled 5-minute stop at Sublette, where the locomotive will take water and the train will be inspected. September 2006.

OPPOSITE PAGE: Sublette nestles into the hillside, well below the top of the mesa. D&RGW section crews lived here year round, and the sheltered location kept them out of the wind. Both photos, September 2006.

Right on schedule, No. 484 and Train No. 215 depart Sublette and head for the deep reaches of Toltec Gorge. The aspens are about to peak for the fall season, the train is close to capacity, and it's a great day for the C&TS. September 2006.

West of Sublette the C&TS line climbs through empty, almost pristine, country that is little changed since the D&RG built through here in 1880. Most of the right of way is only a narrow shelf cut into the hillside. Both photos, September 2006.

As the gorge narrows and deepens the C&TS continues its steady westbound climb. No. 484 and her train pass through ever-larger aspen groves, then reach the long passing siding at Toltec, between Mileposts 310 and 311. Both photos, September 2006.

Mud Tunnel from above, with the remnants of snow from a storm the week before persisting in areas shaded from the sun. Both tunnels on the C&TS are short; this one is just 342 feet long. September 2006.

Long stretches along the C&TS are open to grazing, and the trains occasionally have to slow to chase cattle off the track. As we approach Phantom Curve, this bevy of "the girls" has come out to greet the speeder running ahead of the westbound train. September 2006.

The colorful volcanic formations of Phantom Curve produce an almost surrealistic landscape. The name was applied by D&RG railroaders, who likened shadows cast by their locomotive headlights to phantoms racing away from the trains. No. 484 is near the spot where a snowslide swept three passenger cars of the D&RGW's *San Juan* off the track in February 1948. September 2006.

This train has just left Phantom Curve and is proceeding west. The soft, crumbly volcanic soil here, visible in the light-colored cuts, calls for almost constant maintenance by the C&TS to remove rocks and mud from the track. Both photos, September 2006.

Rock Tunnel is 366 feet long, and was blasted out of solid rock with black powder in 1880. This view of the east portal shows the massive and rugged outcropping that D&RG locating engineers chose to bore through when they couldn't find a way around it.

The west portal from inside. Melting snow from an early storm has formed ice along the rails and a few icicles hanging from the tunnel roof. The Garfield Monument is just out of sight to the right. Both photos, September 2006.

OPPOSITE PAGE: C&TS No. 488 slowly glides out of Rock Tunnel. The masonry retaining wall replaced a temporary wood trestle built here when the line was first pushed through in 1880. September 2006.

Osier – and in time for lunch! There's plenty of snow here at 9,637 ft., even though the aspens have yet to drop their golden leaves. The switch stand just ahead of the locomotive leads to a balloon loop that the C&TS installed for turning engines and trains. September 2006.

New and old structures at Osier. Most important to hungry C&TS passengers is the big yellow dining hall, built in 1989. The brown buildings to the right date to D&RGW days, and include a section house, water tank, and (out of the photo) depot. Stock pens are on the hillside, just ahead of the locomotive. September 2006.

Trains departing Osier to the west cross this large earthen embankment over Osier creek. A corrugated metal culvert carries the creek through the fill, which replaced a wood trestle installed when the line was first built. No. 484 leads a trainload of well-fed passengers toward Cumbres, Chama, and the conclusion of a great day's ride. September 2006.

THE CLIMB FROM OSIER TO CUMBRES maintains the 1.42 percent grade typical of the westbound trip. West of Osier, trains return to the banks of the Rio de los Pinos, following it almost all the way to its headwaters near the pass.

The scenery remains spectacular. Highlights include Cascade trestle, the Los Pinos valley, and Tanglefoot Curve.

With Osier behind, this Chama-bound train is proceeding slowly through the cut at Martinez Point. Crumbling, unstable rock here has proved to be an operational challenge throughout three decades of C&TS operation. September 2006.

133

OPPOSITE PAGE: It's hard to imagine a more photogenic location for a railroad picture than the C&TS steel trestle over Cascade creek. No. 487 is making extra smoke for the photographers on this special freight run. August 2006.

On a brilliant autumn afternoon No. 484 leads a Chama-bound train across Cascade trestle. As with the similar bridge at Lobato, weight restrictions require that westbound helper engines be separated from the road locomotive by enough cars that both would not be on the bridge at the same time. September 2006.

135

West of Cascade the track is again on a narrow shelf, little more than 100 feet above the Rio de los Pinos. September 2006.

Just past Milepost 323 the line makes almost a 90-degree turn to the north and heads deep into the Los Pinos valley. September 2006.

On a showery spring afternoon, No. 487 rolls west over the swollen Rio de los Pinos and the bridge at mile 324.52. The small herd of cows ahead of the engine seems unimpressed by either rain or train. June 2003.

Early in the 2002 season C&TS trains could not run to Osier because of track problems at Martinez Point. Instead, passengers detrained for a box lunch at Los Pinos, and a second locomotive took the train west to Cumbres and Chama. No. 487 has this train ready to head back, while the passengers are off chasing down lunch. May 2002.

137

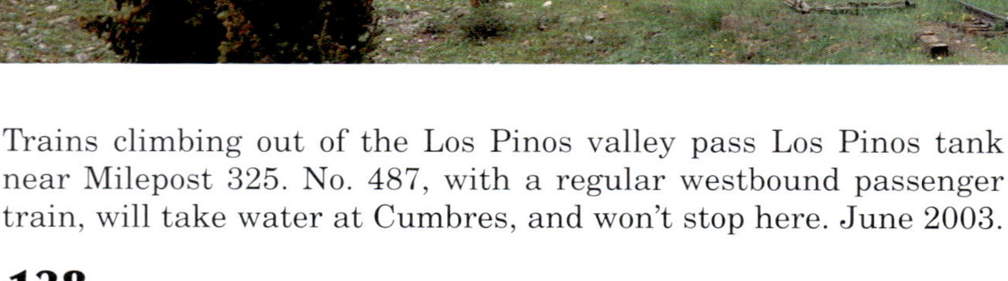

Trains climbing out of the Los Pinos valley pass Los Pinos tank near Milepost 325. No. 487, with a regular westbound passenger train, will take water at Cumbres, and won't stop here. June 2003.

This special freight run is stopping for water at Los Pinos, but more for the photographers than for the locomotive. No. 487's fireman is tugging on the lanyard that opens the water valve. August 2006.

138

Well past the tank, No. 488 pounds upgrade toward the western exit of the Los Pinos valley, just ahead. The line turns left, then right, and passengers will get their first glimpse of the last three miles of the climb to the pass. May 2005.

PRECEDING PAGES: With Highway 17 above and
Cumbres Creek below, No. 484 continues the slow
climb to Cumbres. Behind the train is the entrance
to the vast Toltec Gorge. September 2006.

No. 484 rounds Tanglefoot Curve, still climbing, as trains have
done for a century and a quarter. The sweeping balloon loop
has been a favorite of railroad photographers for the same 125
years. September 2006.

142

The train on the opposite page continues to negotiate Tanglefoot Curve, gaining the last 40 feet of elevation to the yard at Cumbres. In D&RG days the curve, which is both high and exposed to the worst of winter storms, was protected by several wood snow sheds. Most of them burned, some more than once, and the arrival of rotary snow plows in the 1890s changed the road's snow-fighting tactics. The rotaries didn't change the weather, though, and they didn't make railroading on Cumbres much easier. Both photos, September 2006.

Putting on a marvelous show on the last stretch of the climb to the pass, No. 489 reaches the east switch of the passing track. There's only half a mile left before the train makes its 10-minute stop at Cumbres. September 2001.

A westbound switching move along the main at Cumbres, with No. 463 hustling caboose 0503 back to the station area before turing on the wye and heading back to Chama. September 2001.

FOLLOWING PAGES: The pass at last – and it looks like the top of the world from this vantage point! September 2006.

Down the Valley to Chama

Two blasts of the whistle signal the train crew that the descent is about to begin, and No. 487 starts down the four percent grade to Chama. The maximum speed allowed on this section is 12 mph, and there are stretches where speed is restricted to 8 or 10 mph. The 14-mile trip will take an hour and ten minutes. June 2003.

IF THE TRIP UP the four percent helper grade is a battle with gravity, the reverse is true as well. The climb to Cumbres calls for every ounce of tractive effort the locomotives can provide; the return trip to Chama calls for braking power throughout the train and masterful train handling on the part of the engineer.

C&TS trains descend over 2,150 feet in the 14 miles. Modern air brake systems control the descent, and every train crewman knows that safety is paramount. Much of the 10-minute stop at Cumbres is dedicated to inspecting the brake equipment on each and every car in the train.

That said, the views down the valley are breathtaking, and the trip down – without the barking of locomotive exhausts – is surprisingly quiet. In many ways, this seems like a different railroad on the way home to Chama.

A C&TS train tiptoes its way toward the reddish volcanic formations of Windy Point, where the track curves sharply on a narrow shelf perched 250 feet above the valley floor. September 2006.

Rounding Windy Point, No. 484 now heads into the upper valley of Wolf Creek. At its deepest penetration, the line will make a 180-degree turn and head back down the valley, following the creek all the way to Lobato. September 2006.

ABOVE: If the fireman sweats for his paycheck on the trip up the hill, the engineer earns his on the way back down, carefully working engine and train brakes to keep his train under control. No. 484 is inching along toward Coxo, on the high line down from Windy Point.

LEFT: The same train passing Coxo and whistling for the Highway 17 grade crossing. Both photos, September 2006.

OPPOSITE PAGE: There's no need to stop for water at Cresco tank on the downgrade leg of this trip, so No. 487 drifts on by. The locomotive is about to cross a short bridge over the stream that feeds the tank, a tributary of Wolf Creek. May 2005.

Summer wildflowers are in full bloom as No. 484 rolls through Cresco and whistles for the road crossing. The vehicle peeking through the trees above the Cresco signboard is on Highway 17. August 2006.

Departing Cresco amid spring greens, this train passes the yellow telephone booth and crosses a narrow dirt road that connects with Highway 17, off to the left. May 2005.

153

Different season, different colors. As the train rounds a curve west of Cresco, the reds, yellows, and oranges of Gambel's Oak thickets furnish autumn accents.

Farther west, No. 484 rolls the same train through an S curve near Milepost 337. It's not unusual to see bears in this area, especially at this time of year. Both photos, September 2006.

Passengers have a great view across the broadening Wolf Creek valley as their train heads for Chama. The air compressor on No. 484 is doing most of the hard work on this stretch. August 2006.

Dalton is the site of a long-vanished spur track, but it retains its status on the C&TS timetable. The long special freight behind No. 497 is hidden by condensing steam from a boiler blowdown on a cold fall day. October 2002.

Vacation log cabins at Lobo Lodge provide the backdrop as K-37 No. 497 coasts quietly downgrade toward Lobato, where Wolf Creek joins the Rio Chama. May 2002.

One lasting impression of the C&TS is of tiny trains traversing a vast landscape. No. 488 emits only the slightest wisp of coal exhaust as her train rolls downgrade near Milepost 339, just five miles east of Chama. The corrals to the left of the train are part of Lobo Lodge. May 2005.

A long special freight from Cumbres crosses Lobato trestle behind No. 497. Stock cars appear in many photos of D&RGW freights coming off the mountain, because empty stock cars were kept at Cumbres to provide braking power for short trains. October 2002.

In a different season and different weather No. 488 scoots a regular passenger run toward the meadows at Lobato. A westerly breeze is bending both 488's smoke and the aspens behind the train to the right.

The same consist a few minutes later, with No. 488 turning to parallel Highway 17 and enter The Narrows. Soon the train will reach the bottom of the four percent grade, with Chama not far beyond. Both photos, May 2005.

OPPOSITE PAGE: C&TS No. 497 puts on an impressive smoke display for a downgrade run as she enters The Narrows, along Highway 17. May 2001.

With the four percent behind the train, No. 497 drifts down the long stretch of straight track between The Narrows and the final crossing of Highway 17. Next comes the Rio Chama bridge, and home. For the passengers, it's been a long day – and a happy one! May 2001.

Home at last! K-36 locomotive No. 487 leads her train past the monument that is the coaling tower at Chama, just 100 or so yards from the end of today's run. The locomotive and some of the passengers began the day in Antonito; the riders will be bused back there, while No. 484 will turn on the wye and lead tomorrow's eastbound train. September 2006.

Like a faithful hound, speeder No. 101 and the fire patrol putts into Chama a minute or two after the train arrives. May 2005.

Shadows are lengthening as No. 487 slowly clanks into the yard at Chama on a spring afternoon. The crew will complete the day's work by helping the passengers detrain safely, then rearrange the train to prepare for tomorrow. September 2001.

Special Trains and Special Seasons

A triple-headed freight arrives noisily at Cumbres. The C&TS maintains a selection of historic freight equipment in top operating condition for these special events. September 2001.

OPPOSITE PAGE: This special freight being assembled at Chama will include recently restored caboose 0579. August 2006.

EVERY TRAIN ON THE C&TS IS SPECIAL, but some, as the old saying goes, are more special than others. At various times throughout and after the operating season, groups and individuals work with the railroad to produce extra-fare events that differ from normal passenger runs.

Among these events are double- and even tripleheaders, freight trains, photographers' specials, and snow trains.

Often the theme is to reproduce the look of D&RGW operations of half a century ago, with rolling stock relettered for the purpose.

This turns out to be surprisingly easy, because the C&TS is a true operating museum with most of its equipment dating back to the 20th century, and because the territory through which it runs is remote and unchanging.

Another tripleheader with No. 463 in the lead, and about as much smoke and steam as anyone could imagine. October 1999.

K-37 No. 497 is headed for Chama with a photo freight that includes several newly restored stock cars. October 1999.

Doubleheaders on special trains approaching the Coxo crossing, but with the difference in the details. (Above) With "Rio Grande" lettering to evoke D&RGW days, Nos. 463 and 497 roll past freshly fallen snow on the way to Cumbres. October 2002. (Left) With both locomotives wearing Cumbres & Toltec lettering, Nos. 487 and 488 whistle for the Highway 17 crossing and thunder up the four percent grade below Windy Point. August 2006.

OPPOSITE PAGE: There's frost on the ties and steam in the air as No. 463 prepares to couple to a special freight on a frigid Chama morning, after the regular C&TS season. October 2002.

No. 488 backs a restored idler car and pipe gondola toward the rest of a photo freight at the Chama depot, while the passengers look on. The pipe cars recall D&RGW freights of the 1950s and 60s. August 2006.

169

No. 488 belches out a thick plume of exhaust as she covers the last hundred yards of the grade to Cumbres. The load of pipe is headed in the wrong direction (D&RGW hauled pipe west, to the oil and gas fields around Farmington, New Mexico), but the train looks great.

BELOW: The same train is about to round Tanglefoot Curve, en route to Osier. The passenger conversions of boxcars date to the earliest years of C&TS operations, before the road built "proper" passenger trains. Both photos, August 2006.

Our pipe train again, this time turned and putting out extra smoke for photo run-by west of Osier. August 2006.

Acknowledgements

THIS IS A BOOK OF PICTURES, but wonderful people helped make those pictures possible over a period of more than eight years.

First and foremost, sincere thanks are due to all of the C&TS employees, past and present, for their skill and sense of mission in operating the railroad, and for their help with numerous photo shoots.

The same sort of general thanks are due to the Friends of the C&TS. All who love the railroad are forever in their debt for their unrelenting efforts to preserve – and, in recent seasons, operate – the C&TS for the benefit of enthusiasts and visitors – past, present, and future.

Individual thanks go to: Tim Tennant, current CEO of the Friends of C&TS and C&TS General Manager for the 2006 season, who provided extraordinary access to the railroad.

To Frank Turner, President of C&TS Management Corp. for the 2006 season, who served as our motorman and guide for a trip from Antonito to Osier and return in speeder MW-02.

To veteran C&TS Engineer Carlos Llamas, who obligingly stopped and re-started his train for us several times for photos.

To Tom Edison of New Air Helicopters, Durango, Colorado, who was invaluable in arranging for aerial photos, as was pilot John Lee, who expertly piloted his Bell 206 Jet Ranger for us.

To Production Manager Melanie Edbrooke at Four Colour Imports Ltd. for facilitating book production, and General Manager Steve Orf at Four Colour Imports Digital for preparing the pages for printing.

To George H. Drury for expert proofreading, and Charlie Getz, Executive Director of the NGPF, for organizational guidance. To Steve Montano of the C&TS shops and Tony Kassin of the Friends for helping identify photos of the C&TS railroaders.

To Trains Magazine Editor Jim Wrinn for permission to use the C&TS map by Robert Wegner on the front and back end papers.

Two outstanding references on the C&TS deserve special mention. Doris B. Osterwald's superb mile-by-mile guide to the C&TS, *Ticket to Toltec*, is an indispensable reference to the history and geography of the railroad. *The Cumbres & Toltec Scenic Railroad: The Historical Preservation Study*, by Spencer Wilson and Vernon G. Glover, is equally valuable in understanding the museum aspect of the C&TS. Both books are available at this writing.

Finally, my thanks to my friend Bob Hayden, who served as editor, navigator, flight scheduler, art director, and production coordinator. Without Bob this book would never have happened. — *Sam Furukawa*

OPPOSITE PAGE: K-27 No. 463 takes a long drink at the Chama water tank. May 2001.

Headed home toward Chama on a cool spring day, C&TS No. 487 rounds a curve to exit the Los Pinos valley and heads for Tanglefoot Curve and Cumbres. June 2003.

FIRST ESTABLISHED as the Narrow Gauge Trust in 2000, then as the Narrow Gauge Preservation Foundation in 2002, the Foundation is a non-profit public benefit corporation incorporated in the State of California. Its mission is to establish a museum of narrow gauge railroading and models, and to support preservation efforts where possible.

The Foundation supports the efforts, past and current, of numerous existing museums and preservation organizations, and has made grants to several of these organizations to further their work of restoring irreplaceable equipment and structures.

Donations to the Foundation are tax-deductible, and can be either recognized or anonymous. To learn more about making donations to the Foundation, write Executive Director Charlie Getz at:

Narrow Gauge Preservation Foundation
P.O. Box 1073
San Carlos, CA 94070-1073

Steam and smoke almost obscure a stand of golden aspens as a C&TS doubleheader thunders up the grade toward Cumbres. October 2002.

Index

Inoperable and in need of heavy repairs since 2002, C&TS No. 463 snoozes peacefully in the Antonito shops, waiting for better days. September 2006.

FOLLOWING PAGE: After the end of the 2002 season, K-27 No. 463 leads an empty passenger train eastward to Antonito through fresh snow. October 2002.

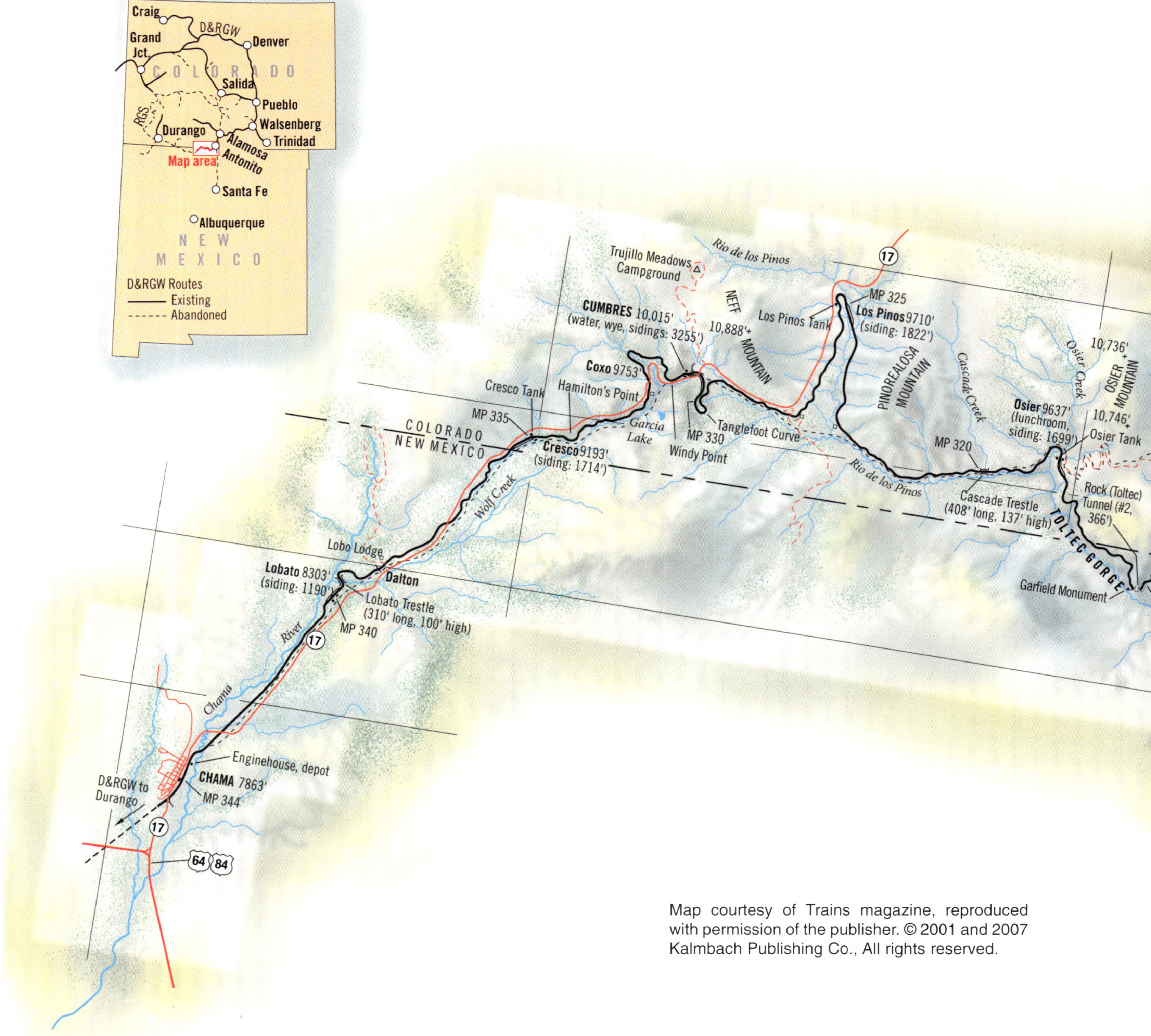

Map courtesy of Trains magazine, reproduced with permission of the publisher. © 2001 and 2007 Kalmbach Publishing Co., All rights reserved.